# YOUR KNOWLEDGE HAS VALUE

- We will publish your bachelor's and
  master's thesis, essays and papers

- Your own eBook and book -
  sold worldwide in all relevant shops

- Earn money with each sale

Upload your text at www.GRIN.com
and publish for free

David Brückner

# Deceleration of a Conducting Disc with Eddy Currents

## A Practical Investigation

GRIN Verlag

**Bibliografische Information der Deutschen Nationalbibliothek:**

Die Deutsche Bibliothek verzeichnet diese Publikation in der Deutschen National-
bibliografie; detaillierte bibliografische Daten sind im Internet über http://dnb.d-
nb.de/ abrufbar.



**Imprint:**

Copyright © 2012 GRIN Verlag GmbH
Druck und Bindung: Books on Demand GmbH, Norderstedt Germany
ISBN: 978-3-656-34877-1

**This book at GRIN:**

http://www.grin.com/en/e-book/207369/deceleration-of-a-conducting-disc-with-
eddy-currents

# DECELERATION OF A DISC WITH EDDY CURRENTS

David Brückner, UVI

Practical Investigation Coursework 2011

Lancing College

# TABLE OF CONTENT

# 1. Introduction and Theory

When an electric conductor is exposed to a changing magnetic field, there is a force on each electron in the conductor, namely **F** = **B**e**v**, where **B** is the magnetic field strength and **v** the velocity of the particle. This causes an overall induced electromotive force which is equal to the rate of change of flux linkage, $\phi$ = **BAN** $\sin\theta$, with respect to time (Faraday's Law). The emf is such that it opposes the direction of change of flux (Lenz's Law):

$$\varepsilon = -\frac{d\Phi}{dt}$$

The induced emf causes a so-called eddy current to flow, which again has a magnetic field. The magnetic field causing the induction and the field of the induces current the interfere. In situations where either the magnet or the conductor is rotating, the interference can cause a change in velocity.

This can happen in two different ways. In the first scenario, the disc rotates and a stationary magnet is fixed next to the disc. So, relative to the disc, the magnet is moving and hence there is a change in flux and as the magnetic field is at 90° to the direction of motion of the conductor, the flux linkage is maximised. This change of flux then induces an emf across the disc, and a current flows in the direction opposing the direction of rotation. This magnetic field and that of the magnet then interfere forming a retarding effect.

Otherwise, a disc can follow a spinning magnet, or a magnet fixed to an axis allowing free rotation follows a spinning conductor disc. Here, the magnetic field induced in the conductor is attracted by the other field and hence, there is an acceleration effect.

## 2. Aim

The aim of this coursework is to examine the relationship between the deceleration effect of the eddy currents and the initial velocity of the disc as well as the strength of the magnetic field. Additionally, I planned to compare this to the acceleration effect that can be produced using a different set-up in which a freely rotating magnet is following a motorised conductor disc.

# 3. Experimental Procedure

## 3.1 The Main Experiment

The basic idea of the first experiment was to have an aluminium disc that rotates through the field of an electromagnet and is then decelerated. The software used for tracking the data given out by the light gates was EasySense and it didn't allow measurements with a decent sampling rate for longer than 5 minutes. Therefore, I had to find combination of voltage and number of turns of the coil in the electromagnet that produced a detectable effect in that time interval. This was a voltage between 5 to 15 volts and a coil number of 500.

Figure 1, 2: Arrangement of the electromagnet

In order to accelerate the disc reliably in the first place, a motor has been connected via an elastic band to the disc. The disc was then accelerated for a given amount of time until it was disconnected from the motor, in order to let it rotate freely, and at the same instant, the electromagnet was switched on to start the deceleration. The potential difference across the motor was constantly 3.00 ± 0.10 V.

Figure 3: Connection to the motor

The angular velocity was to be measured using light gates. To get these working well, I had to try out several different arrangements. My first approach was to have two light gates and a piece of cardboard of known width stuck to the edge of the disc.

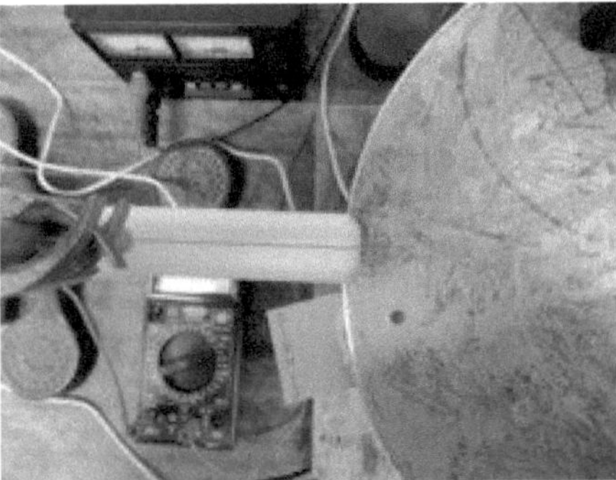

Figure 4, 5: First approach to the measurement of the angular velocity of the disc

However, this method was not satisfactory as the sampling rate of the light sensors too low to detect the cardboard. It could often pass through the gates without being captured. Low spurious frequencies (aliases) were given out which is typical for sampling with too low a sampling rate. The Sampling-Theorem by C.E. Shannon states that the sampling frequency should be at least twice as high as the frequency of the signal to be measured.[1] This is the reason why CD signals have a frequency of 44,100 Hz as the maximum frequency that can be heard by the human ear is 20,000 Hz. This problem with missing passes still occured in my final set-up, but only at the very high speeds I used in some additional experiments.

To avoid this, I attached a semi-circular piece of cardboard to the edge of the disc which covered exactly one half of the circumference of the disc. I also abandoned the second light gate, as it made the data very confusing. Hence, the light gates would always measure the time taken to do half a revolution which makes it easy to calculate the angular velocity.

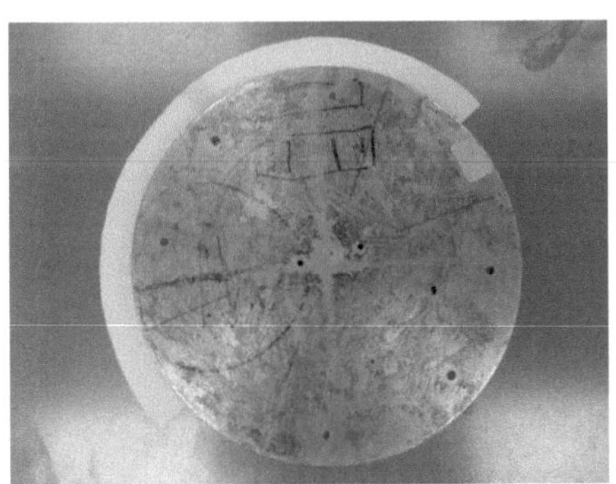
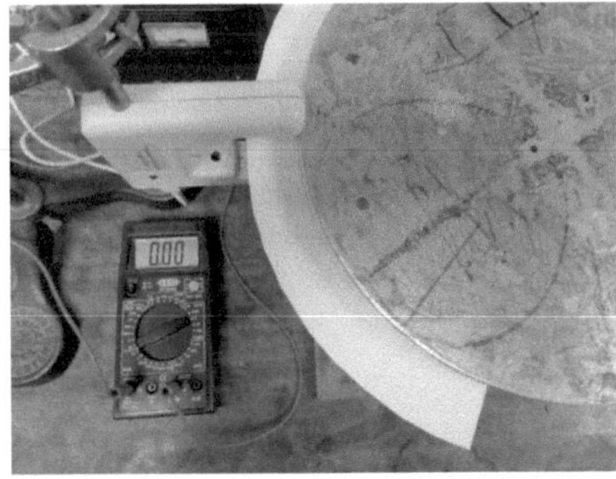

Figure 6, 7: Second approach to the measurement of the angular velocity of the disc

---

[1] From: *An Introduction to the Sampling Theorem*, http://www.ti.com/lit/an/snaa079c/snaa079c.pdf, page 2

This layout now allowed me to do the measurements I wished to do. I decided to repeat every measurement once. The plan was as follows:

|  | 10s acceleration | 20s acceleration | 30s acceleration |
| --- | --- | --- | --- |
| Electromagnet at 5.0 ± 0.1 V | 1.1.1 and 1.1.2 | 1.2.1 and 1.2.2 | 1.3.1 and 1.3.2 |
| Electromagnet at 10.0 ± 0.1 V | 2.1.1 and 2.1.2 | 2.2.1 and 2.2.2 | 2.3.1 and 2.3.2 |
| Electromagnet at 15.0 ± 0.2 V | 3.1.1 and 3.1.2 | 3.2.1 and 3.2.2 | 3.3.1 and 3.3.2 |

Additionally, I did two runs without the magnet, in order to find the deceleration due to friction so that I could subtract this from the total deceleration in the other experiments in order to find the deceleration due to the induction.

## 3.2 Additional Experiments

Moreover, I used the EasySense magnetic field sensor to not only measure the magnetic field strength produced by the magnet at different voltages, but also to investigate the magnitude to the fluctuations due to the heat produced in the coil.

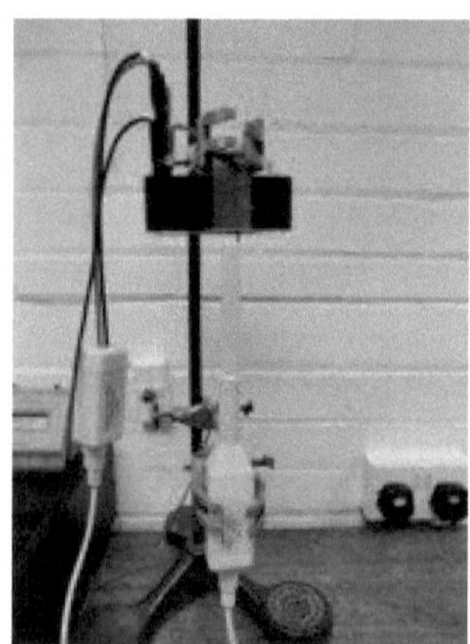

Figure 8, 9: Measuring the magnetic field strength

I measured the magnetic field strength over a period of 2 minutes and at the three different voltages used. Moreover, I repeated each measurement at three different distances from the iron core. The thickness of the aluminium layer on the disc is 0.6 cm and the spacing between the disc and the iron core in the experiments was 0.3 cm. To find the force at the top, in the middle and on the bottom of the disc, I therefore measured it at 0.3, 0.6 and 0.9 cm distance, as shown in the diagram. This was only to get an idea about how much the magnetising force varies with separation.

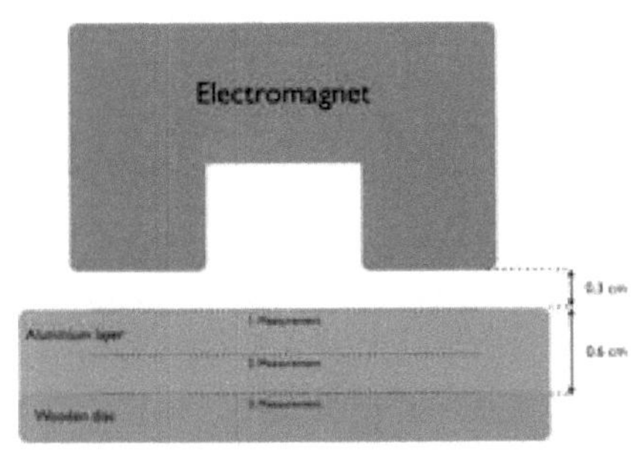

Figure 10, 11: Considering the thickness of the aluminium layer in the context of changing magnetic field strength.

The field to be measured was an axial field. I used both the ±10 mT and the ±100 mT sensor, accordingly to the order of magnitude of the field strength as I started using the ±10 mT sensor and only the later measurements required a greater range.

The results on the next page show the mean of the axial field in mT and the voltage in V over the period of 2 minutes.

|  | 0.3 cm | | | 0.6 cm | | | 0.9 cm | | |
|---|---|---|---|---|---|---|---|---|---|
|  | Axial Field / mT | Voltage / V | Current / A | Axial Field / mT | Voltage / V | Current / A | Axial Field / mT | Voltage / V | Current / A |
| 5V | 8.00 | 4.89 | 1.00 | 4.34 | 4.99 | 1.00 | 3.04 | 4.92 | 1.00 |
| 10V | 14.90 | 10.10 | 1.70 | 8.62 | 10.07 | 1.70 | 6.09 | 10.04 | 1.70 |
| 15V | 22.07 | 14.54 | 2.30 | 11.55 | 14.67 | 2.30 | 8.95 | 14.71 | 2.30 |

The values for the voltage in the left-had column show the approximate values. The uncertainty in these values is as follows:

I.    $5.00 \pm 0.10\,V$

II.   $10.00 \pm 0.10\,V$

III.  $15.00 \pm 0.20\,V$

The power supply got less reliable at potential differences above ~12 V. As I connected an EasySense voltmeter across the magnet, I could find possible aberrations from the anticipated value and put it into consideration.

## 3.3 Acceleration of a disc

In order to accelerate a disc, the same physical phenomenon is used. Here, either a conducting disc follows a spinning magnet, or a magnet fixed to an axis allowing free rotation follows a spinning conductor disc (see Introduction). The spinning magnet is the better-known example, so I decided to have a go at the second one. The Physics Departement had an aluminium disc mounted to a motor, so I only had to find a way of arranging the magnets. I chose to fix two very strong coin-sized magnets to a disc by sticking two flat pieces of metal to the back and subsequently glue the magnet to the disc. The disc was fixed firmly to a screw to which a plastic ring was loosely attached on the back-side and a screw nut prevented it from slipping down. Then, the ring could be clamped to a stand, allowing free rotation. The two magnets were placed on one diameter and equally far away from the centre of the disc (2.25 cm).

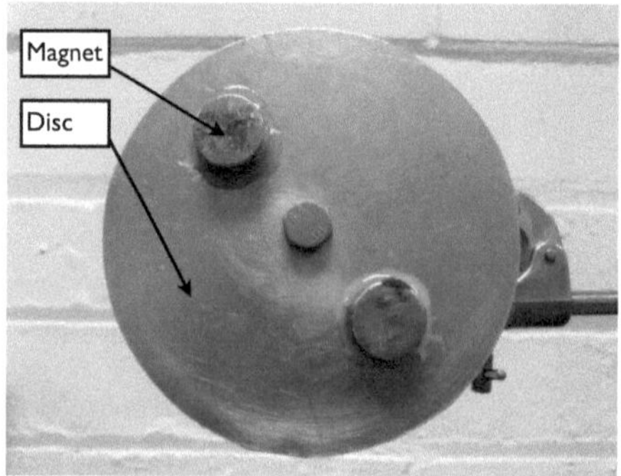


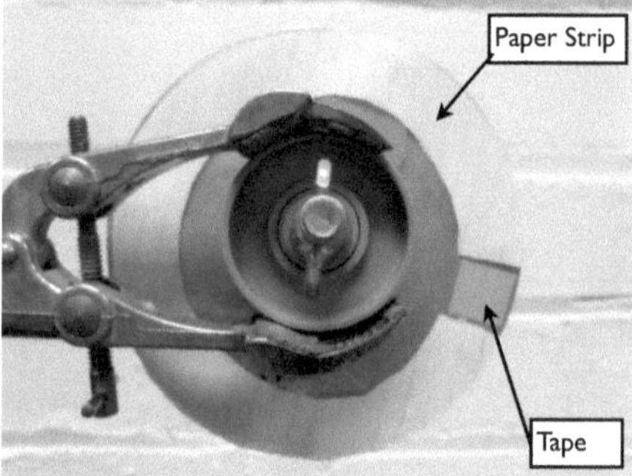


Figure 12, 13: Arranging the freely rotating magnet.

This apparatus was then put face to face with the spinning disc:

Figure 14, 15: Arranging the discs.

As the strength of the magnets could not be changed in this experiment, the aim was to compare the speeds of the two discs rather than speed and magnetic field strength. This time, I used the laser pistol for either disc which gave the angular velocity in rotations per minute (RPM). As they had a "hold"-button, I could press it simultaneously and hence take measurements of the velocity every 20 seconds. Reflecting tape had to be attached to the two discs. During the first measurements, the angular velocity of the magnets was about three times higher than that of the the motor-driven disc, which had to be an experimental error. I concluded, that the shiny surface of the disc to which the magnets were attached probably confused the instrument. Therefore, I fitted white paper along the circumference of the reflecting tape (see figures 15 and 18) and subsequently obtained reasonable results.

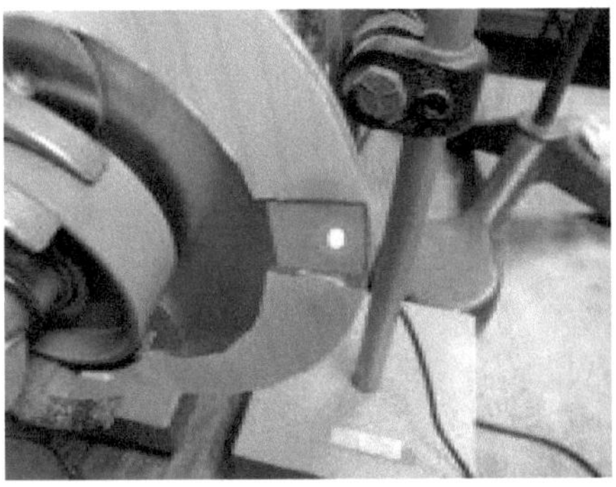

Figure 16, 17: Using reflecting tape.

In order to have two different velocities of the motorised disc, I conducted two experiments, one with a voltage of 2V across the motor, the other with 4V. I recorded the velocities for 2 minutes to observe the acceleration effect.

## 3.4 Risk assessment

This investigation did not hold many risks, the only thing I had to look out for was the heat produced in the coil, as the insulation might start melting if it is switched on for to long. I generally did not keep the magnet switched on for very long periods of time and there was always plenty of time to cool down while I was preparing the next experiment. To check this, I measured the temperature of the magnet using the infrared camera after having used it in a typical experiment.

Figure 18, 19: Measuring the temperature developed in the electromagnet during the experiment.

Due to interference with the cables and the clamp on top, I measured from the bottom. To center the camera to the right point, I used a mirror underneath it. The measured temperature after, having used the magnet for a considerably long time, was 62 °C which according to my physics teacher was in the acceptable range, and would not cause any harm.

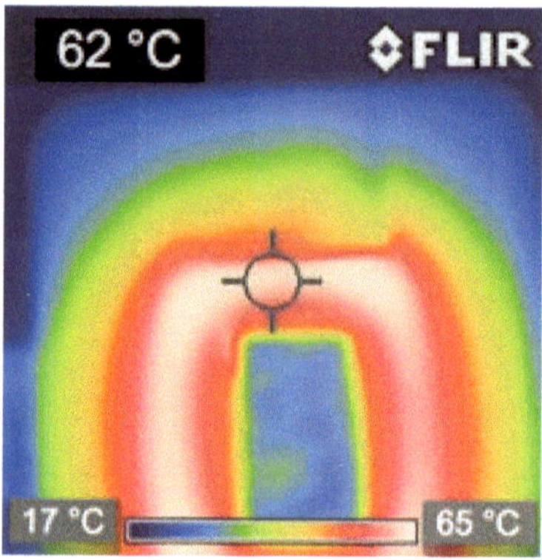

Figure 20: Measurement of the temperature of the electromagnet.

# 4. Results

## 4.1 Extracting from the raw data

The light gates took a measurement of the relative intensity every 0.05 sec. The signal was either "100%" or "3.2%" depending on wether the cardboard was just passing through or not. The software used, EasySense, was unable to perform even the simplest calculations so this raw data had to be further processed in Excel.

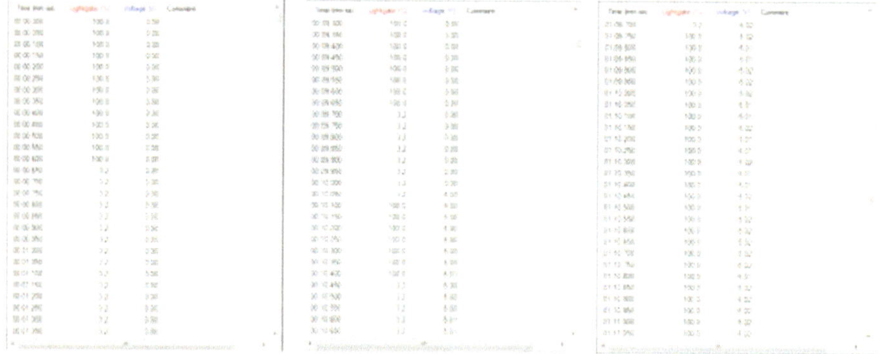

Figure 21, 22, 23: Raw data at the beginning (23), somewhere in the middle (24) and at the end (25) of the measurement.

$$\omega = \frac{2\pi}{T}$$

Angular velocity is 2π divided by the time taken to travel 2π radians, the period. As the blocks of equal intensity (see figure 24) represent half a revolution, the period is the time elapsed for two of those blocks (one "3.2%" and one "100%" block). Doing this in Excel was not at all that easy.

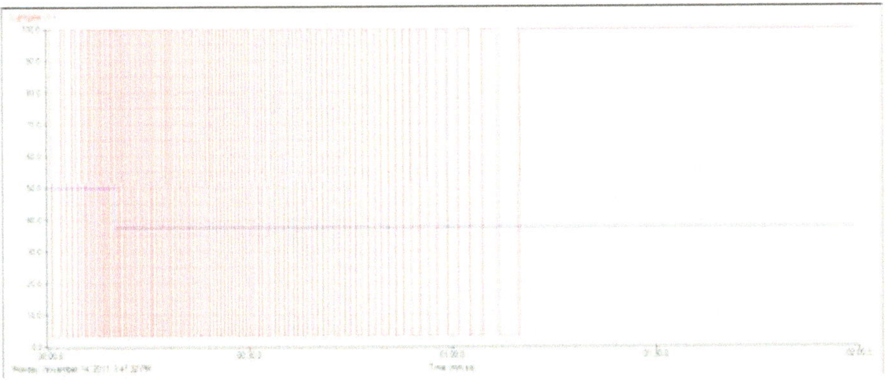

Figure 24: Graph produced by EasySense for experiment 1.1.1. The red curve represents the light gate signal, the blue one the voltage.

The following example is taken from experiment 1.1.1:

|  | A | B | C | E | F | G | H | I |
|---|---|---|---|---|---|---|---|---|
|  | Time | Lightgate | Voltage | Lightgate rounded | Time | Effective Time | | Angular velocity |
|  | s | % | V | % | s | s | s | rad s⁻¹ |
| 273 | 7.849999905 | 100 | 0 | 100 | 6.8 | 6.8 | 0 | |
| 274 | 7.900000095 | 100 | 0 | 100 | 6.9 | 6.8 | 0 | |
| 275 | 7.949999809 | 100 | 0 | 100 | 6.9 | 6.8 | 0 | |
| 276 | 8 | 100 | 0 | 100 | 7 | 6.8 | 0 | |
| 277 | 8.050000191 | 100 | 0 | 100 | 7 | 6.8 | 0 | |
| 278 | 8.100000381 | 100 | 0 | 100 | 7.1 | 6.8 | 0 | |
| 279 | 8.149999619 | 100 | 0 | 100 | 7.1 | 6.8 | 0 | |
| 280 | 8.199999809 | 3.20000005 | 0 | 0 | 7.1 | 6.8 | 0 | |
| 281 | 8.25 | 3.20000005 | 0 | 0 | 7.2 | 6.8 | 0 | |
| 282 | 8.300000191 | 3.20000005 | 0 | 0 | 7.3 | 6.8 | 0 | |
| 283 | 8.350000381 | 3.20000005 | 0 | 0 | 7.3 | 6.8 | 0 | |
| 284 | 8.399999619 | 3.20000005 | 0 | 0 | 7.3 | 6.8 | 0 | |
| 285 | 8.449999809 | 3.20000005 | 0 | 0 | 7.4 | 6.8 | 0 | |
| 286 | 8.5 | 3.20000005 | 0 | 0 | 7.5 | 6.8 | 0 | |
| 287 | 8.550000191 | 100 | 0 | 100 | 7.5 | 7.5 | 0.7 | 8.97597901 |

Example 1: Raw data of one period in experiment 1.1.1

The algorithm used to obtain the velocity works as follows (here outlined using the example of row 273):

I.     Column E: The intensity value from the light barrier is to be rounded to either 100 or 0. This is done by **=ROUND(B273,-1)**.

II.    Column F: The first rotation in each experiment had to be excluded from all calculations, as the starting point along the circumference were different every time. Therefore, I had to adjust the time parameter by **=ROUND(A273-A$25,1)**. 25 is here the first row that is included in the calculation, $ means that 25 is fixed down the column whereas 273 is a parameter running through.

III.   Column G: This is the first more sophisticated operator: it picks out the rows where the the light gate signal changes from 100 to 0, but not vice versa in order to treat two blocks as a whole revolution. It puts the current time parameter down if there is a change and if there is not, the value from the previous cell is recorded. This is **=IF(AND(E273=100,E236=0),F237,G236)**.

IV.    Column H: Now the time elapsed has to be printed. I did this by setting the difference in time if this is greater zero and otherwise zero. As in column G the times only differ when a full revolution is performed, this will then give the period in those cells and zero otherwise. This is done by **=IF(G237-G236>0,G237-G236,0)**.

V.     Column I: Having obtained the period, the angular velocity can be calculated as outlined earlier, namely by **=IF(H237>0,2*PI()/H237,"")** which will fill in the velocity at the end of the data belonging to one rotation and leave the cell blank otherwise.

The data in the table is not rounded to a sensible amount of significant figures as this is still raw data and this will be done in the next step. All these calculations had to be fully automatic as there were $2\times3\times4 = 24$ different experiments, each containing roughly $120\div0.05 = 2400$ measurements, all together 57,600 measurements. This amount of data could of course not be managed by hand.

In the next step, I selected all the rows that contained a value of the angular velocity (in the example row 287) and pasted them into a new table which could then be used to create a graph, plotting angular velocity versus time. Only the values measured after the magnet was switched on where included, in order to be able to fit a regression line in order to find the gradient, the deceleration.

The light gates took a measurement every 0.05 seconds. Therefore, there was an uncertainty of $\pm0.025$s for the half-period, as this was what the light barriers really measured as the cardboard covered half of the perimeter. The standard error in the period was hence $\pm0.05$s, resulting in an error in the angular velocity that I calculated as follows:

$$\text{Positive error} = \frac{2\pi}{T - 0.05} - \omega_{\text{Measured}} \qquad \text{Negative error} = \omega_{\text{Measured}} - \frac{2\pi}{T + 0.05}$$

These values were calculated individually for each piece of data and were included in the diagram in the form of error bars.

## 4.2 Accounting for friction and other energy losses

In order to be able to differentiate the deceleration caused by the induction and friction, air resistance, etc., I undertook three measurements without using the magnet with the three different initial speeds used in the other experiments.

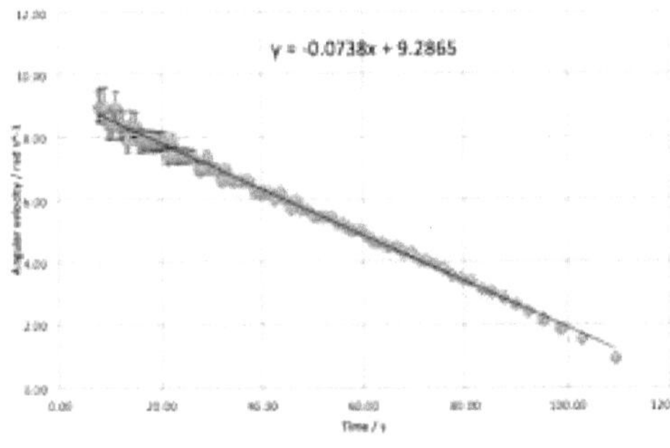

Diagram 1: Testing for friction with 10s acceleration.

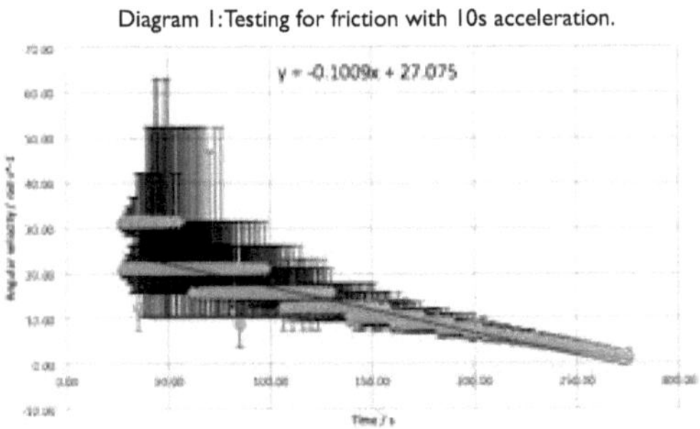

Diagram 2: Testing for friction with 20s acceleration.

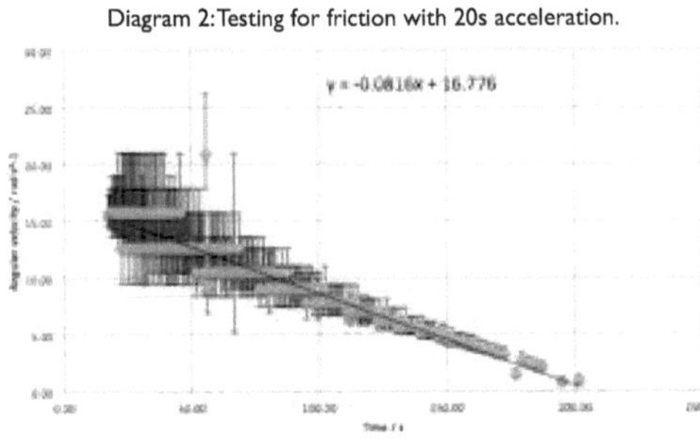

Diagram 3: Testing for friction with 30s acceleration.

The three graphs show the angular velocity with respect to time. Only the data recorded just after the motor was disconnected from the disc is plotted, otherwise, a regression line could not be fitted. The error bars are calculated in the fashion described in the previous chapter. In diagram 2 and 3 there are very large uncertainties as well as "jumps" in the velocity. This is because very high speeds had to be recorded here were the standard error ±0.05s in the period makes a lot of difference; the successive plateaus that can be observed are only due to this difference as the all lie in each others error bar range.

The graphs all have a reasonably linear shape, therefore I assumed that the deceleration was constant as the velocity changes, in order to simplify further calculations. I took the value of the acceleration to be the gradient of the regression line.

I used the value of the deceleration due to friction in the main experiments to adjust the tabulated angular velocity so that only the deceleration due to induction would affect the data.

This was done by subtracting the deceleration due to friction multiplied by the time difference between the corresponding reading and the previous reading from the current angular velocity measured:

$$\omega_{\text{Corrected}} = \omega_{\text{Measured}} - a \times (t_{\text{Corresponding}} - t_{\text{Previous}})$$

As $a$ is always negative, the corrected angular velocity will be greater than the measured velocity, and approximately by the amount that is lost due to friction. The results can be seen on the following page.

| | 10s acceleration | 20s acceleration | 30s acceleration |
|---|---|---|---|
| Deceleration due to friction / rad s⁻¹ | -0.074 | -0.10 | -0.082 |

The following example shows the data extracted from the record of experiment 1.1.1. The row marked in orange is where the electromagnet was switched on, the data before is not plotted on the diagram.

| Time | Lightgate | Voltage | | Lightgate ro | Time | Effective Time | | Angular velocity | MAX | MIN | Angular velocity |
|---|---|---|---|---|---|---|---|---|---|---|---|
| s | % | V | | % | s | s | s | rad s^-1 | rad s-1 | rad s^-1 | rad s^-1 |
| 3.75 | 100 | 0.00 | | 100 | 1.70 | 1.70 | 1.70 | 3.70 | 0.11 | 0.11 | |
| 5.00 | 100 | 0.00 | | 100 | 2.95 | 2.95 | 1.25 | 5.03 | 0.21 | 0.19 | |
| 6.05 | 100 | 0.00 | | 100 | 4.00 | 4.00 | 1.05 | 5.98 | 0.30 | 0.27 | |
| 6.95 | 100 | 0.00 | | 100 | 4.90 | 4.90 | 0.90 | 6.98 | 0.41 | 0.37 | |
| 7.80 | 100 | 0.00 | | 100 | 5.75 | 5.75 | 0.85 | 7.39 | 0.46 | 0.41 | |
| 8.55 | 100 | 0.00 | | 100 | 6.50 | 6.50 | 0.75 | 8.38 | 0.60 | 0.52 | |
| 9.30 | 100 | 0.00 | | 100 | 7.25 | 7.25 | 0.75 | 8.38 | 0.60 | 0.52 | |
| 10.10 | 100 | 4.99 | | 100 | 8.05 | 8.05 | 0.80 | 7.85 | 0.52 | 0.46 | 7.85 |
| 10.85 | 100 | -5.00 | | 100 | 8.80 | 8.80 | 0.75 | 8.38 | 0.60 | 0.52 | 8.43 |
| 11.60 | 100 | -5.00 | | 100 | 9.55 | 9.55 | 0.75 | 8.38 | 0.60 | 0.52 | 8.43 |
| 12.40 | 100 | -5.00 | | 100 | 10.35 | 10.35 | 0.80 | 7.85 | 0.52 | 0.46 | 7.91 |
| 13.20 | 100 | -5.00 | | 100 | 11.15 | 11.15 | 0.80 | 7.85 | 0.52 | 0.46 | 7.91 |
| 14.05 | 100 | -5.00 | | 100 | 12.00 | 12.00 | 0.85 | 7.39 | 0.46 | 0.41 | 7.45 |
| 14.85 | 100 | -5.00 | | 100 | 12.80 | 12.80 | 0.80 | 7.85 | 0.52 | 0.46 | 7.91 |
| 15.70 | 100 | -5.00 | | 100 | 13.65 | 13.65 | 0.85 | 7.39 | 0.46 | 0.41 | 7.45 |
| 16.60 | 100 | -5.00 | | 100 | 14.55 | 14.55 | 0.90 | 6.98 | 0.41 | 0.37 | 7.06 |
| 17.45 | 100 | -5.00 | | 100 | 15.40 | 15.40 | 0.85 | 7.39 | 0.46 | 0.41 | 7.45 |
| 18.35 | 100 | -5.00 | | 100 | 16.30 | 16.30 | 0.90 | 6.98 | 0.41 | 0.37 | 7.06 |
| 19.25 | 100 | -5.01 | | 100 | 17.20 | 17.20 | 0.90 | 6.98 | 0.41 | 0.37 | 7.06 |
| 20.20 | 100 | -5.00 | | 100 | 18.15 | 18.15 | 0.95 | 6.61 | 0.37 | 0.33 | 6.68 |
| 21.15 | 100 | -5.00 | | 100 | 19.10 | 19.10 | 0.95 | 6.61 | 0.37 | 0.33 | 6.68 |
| 22.10 | 100 | -5.00 | | 100 | 20.05 | 20.05 | 0.95 | 6.61 | 0.37 | 0.33 | 6.68 |
| 23.10 | 100 | -5.01 | | 100 | 21.05 | 21.05 | 1.00 | 6.28 | 0.33 | 0.30 | 6.36 |
| 24.10 | 100 | -5.00 | | 100 | 22.05 | 22.05 | 1.00 | 6.28 | 0.33 | 0.30 | 6.36 |
| 25.15 | 100 | -5.00 | | 100 | 23.10 | 23.10 | 1.05 | 5.98 | 0.30 | 0.27 | 6.06 |
| 26.20 | 100 | -5.00 | | 100 | 24.15 | 24.15 | 1.05 | 5.98 | 0.30 | 0.27 | 6.06 |
| 27.30 | 100 | -5.00 | | 100 | 25.25 | 25.25 | 1.10 | 5.71 | 0.27 | 0.25 | 5.79 |
| 28.40 | 100 | -5.00 | | 100 | 26.35 | 26.35 | 1.10 | 5.71 | 0.27 | 0.25 | 5.79 |
| 29.55 | 100 | -5.01 | | 100 | 27.50 | 27.50 | 1.15 | 5.46 | 0.25 | 0.23 | 5.55 |
| 30.70 | 100 | -5.00 | | 100 | 28.65 | 28.65 | 1.15 | 5.46 | 0.25 | 0.23 | 5.55 |
| 31.90 | 89.4 | -5.00 | | 90 | 33.70 | 33.70 | 1.30 | 4.83 | 0.19 | 0.18 | 5.21 |
| 33.15 | 100 | -5.00 | | 100 | 31.10 | 31.10 | 1.25 | 5.03 | 0.21 | 0.19 | 4.83 |
| 34.45 | 100 | -5.00 | | 100 | 32.40 | 32.40 | 1.30 | 4.83 | 0.19 | 0.18 | 4.93 |
| 35.75 | 92.6 | -5.00 | | 90 | 33.70 | 33.70 | 1.30 | 4.83 | 0.19 | 0.18 | 4.93 |
| 37.15 | 100 | -5.00 | | 100 | 35.10 | 35.10 | 1.40 | 4.49 | 0.17 | 0.15 | 4.59 |
| 38.55 | 100 | -5.00 | | 100 | 36.50 | 36.50 | 1.40 | 4.49 | 0.17 | 0.15 | 4.59 |
| 40.05 | 100 | -5.00 | | 100 | 38.00 | 38.00 | 1.50 | 4.19 | 0.14 | 0.14 | 4.30 |
| 41.60 | 100 | -5.00 | | 100 | 39.55 | 39.55 | 1.55 | 4.05 | 0.14 | 0.13 | 4.17 |
| 43.20 | 100 | -5.00 | | 100 | 41.15 | 41.15 | 1.60 | 3.93 | 0.13 | 0.12 | 4.05 |
| 44.90 | 100 | -5.00 | | 100 | 42.85 | 42.85 | 1.70 | 3.70 | 0.11 | 0.11 | 3.82 |
| 46.70 | 100 | -5.00 | | 100 | 44.65 | 44.65 | 1.80 | 3.49 | 0.10 | 0.09 | 3.62 |
| 48.60 | 100 | -5.00 | | 100 | 46.55 | 46.55 | 1.90 | 3.31 | 0.09 | 0.08 | 3.45 |
| 50.60 | 100 | -5.01 | | 100 | 48.55 | 48.55 | 2.00 | 3.14 | 0.08 | 0.08 | 3.29 |
| 52.80 | 100 | -5.00 | | 100 | 50.75 | 50.75 | 2.20 | 2.86 | 0.07 | 0.06 | 3.02 |
| 55.15 | 100 | -5.01 | | 100 | 53.10 | 53.10 | 2.35 | 2.67 | 0.06 | 0.06 | 2.85 |
| 57.80 | 100 | -5.00 | | 100 | 55.75 | 55.75 | 2.65 | 2.37 | 0.05 | 0.04 | 2.57 |
| 60.85 | 100 | -5.01 | | 100 | 58.80 | 58.80 | 3.05 | 2.06 | 0.03 | 0.03 | 2.29 |
| 64.50 | 100 | -5.00 | | 100 | 62.45 | 62.45 | 3.65 | 1.72 | 0.02 | 0.02 | 1.99 |
| 69.75 | 100 | -5.01 | | 100 | 67.70 | 67.70 | 5.25 | 1.20 | 0.01 | 0.01 | 1.58 |

Example 2: Processed data from experiment 1.1.1

The first nine columns are the results copied from the raw data. The 10th and 11th column show the positive and negative error respectively, calculated as outlined in section 4.1. The last column contains the the corrected angular velocity (see p. 14). The first value in this column, in the orange row is not corrected as no deceleration due to friction will have influenced it as it is measured at the instant when the motor is disconnected from the disc.

## 4.3 Fitting a regression line

On my diagrams, I plotted the corrected angular velocity against the time. To fit a regression line, the relationship between $\omega$ and the time had to be found. The curves have a shape which suggests an exponential relationship:

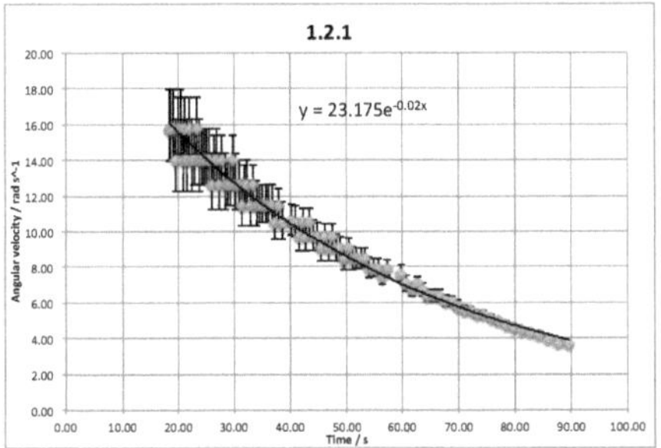


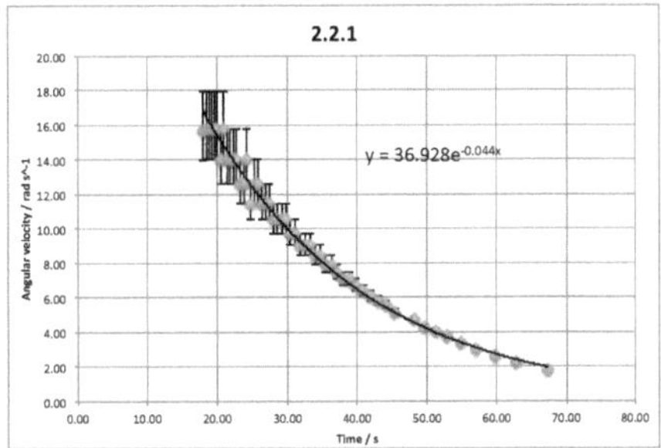


**Diagram 4, 5:** Plot of the corrected angular velocity against the time in seconds from experiment 1.2.1 (acceleration for 20s and electromagnet at 5V) and experiment 2.2.1 (acceleration for 20s and electromagnet at 10V) respectively.

In order to verify this idea, I plotted the natural logarithm of the angular velocity against time. If the relationship was exponential, this should give a straight line with negative gradient, as:

$$\text{If } \omega = Ae^{-\lambda t}, \text{ then } \ln(\omega) = -\lambda t + \ln(A).$$

I have done this for every graph, but here I will only display the ones corresponding to the examples above.

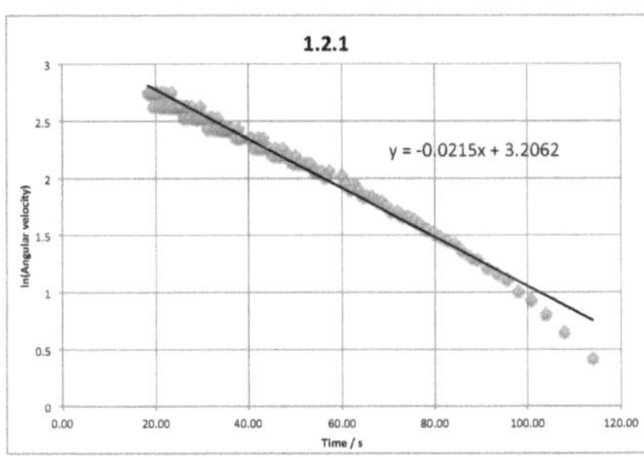


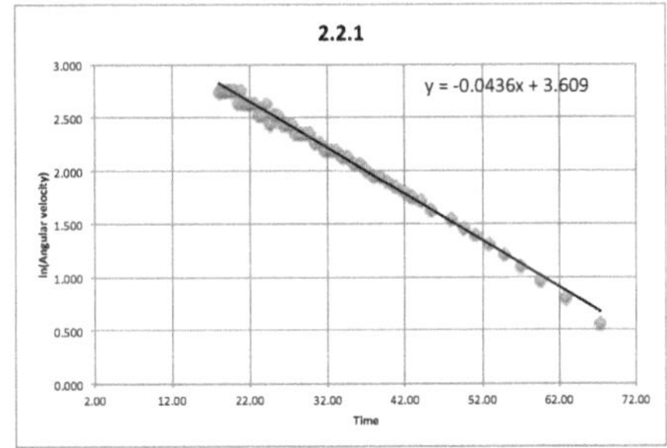


**Diagram 6, 7:** Plot of the natural logarithm of the corrected angular velocity against the time in seconds from experiment 1.2.1 and 2.2.1 respectively.

The graph 1.1.1 has a fairly linear shape, however the points fade at the end. The graph for 1.2.1 however is very straight, and here the values at the end only fade very slightly. The equation of the regression line here is strong evidence that the relationship is indeed exponential: The regression line fitted to diagram 5 gives $\lambda = -0.044$ and $A = 36.928$, the regression line in diagram 7 has values $\lambda = -0.0436$ and $\ln(A) = 3.609$, giving $A = 36.929$ (see equation above for reference). These values are very close and confirm the assumption that the relationship is exponential.

In experiment 1.1.1, the curve looks more linear than exponential and the points on the ln-graph fade near the end of the experiment:

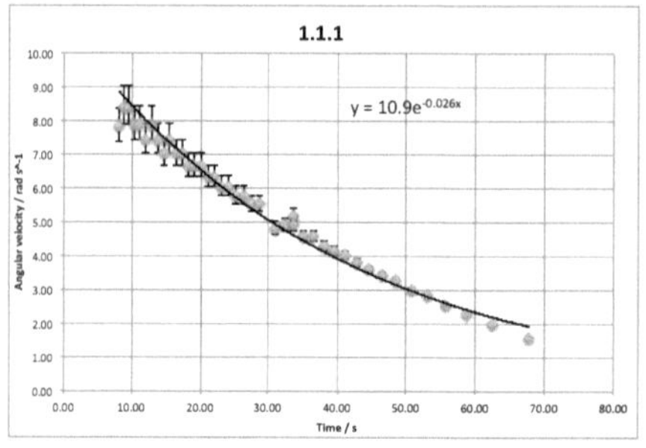


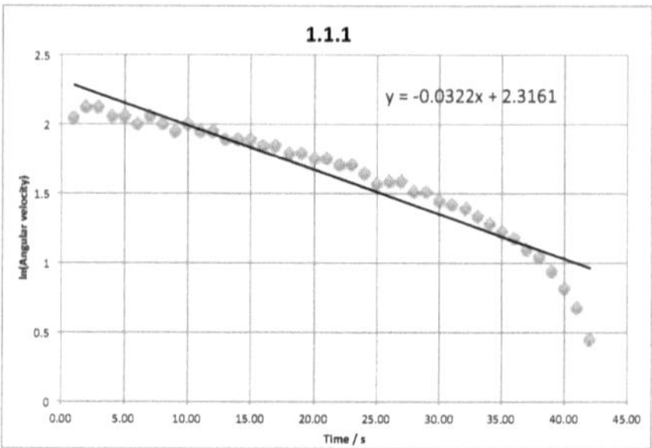


Diagram 8, 9: Graphs for experiment 1.1.1 showing angular velocity and the natural log of the angular velocity versus time.

A possible explanation for the discontinuities and the unexpectedly low values at t > 40s is that friction has a greater effect at very low speeds, and is hence not linear as assumed for the calculation of the corrected angular velocity.

## 4.4 Main Experiment

For every experiment, I plotted the graph of angular velocity against time in order to find the decaying factor of the exponential, which I will refer to as $\lambda$. This factor is a measure of the rate of decay as:

$$\frac{d}{dt}(e^{-\lambda t}) = -\lambda e^{-\lambda t} \quad \text{and} \quad t_{0.5} = \frac{\ln(2)}{\lambda}$$

So the magnitude of the deceleration is proportional to $\lambda$, meaning that the greater $\lambda$, the quicker the decay and the half-life $t_{0.5}$ is inversely proportional to $\lambda$, implying that the greater $\lambda$, the shorter it takes to halve the amount and hence the quicker the decay.

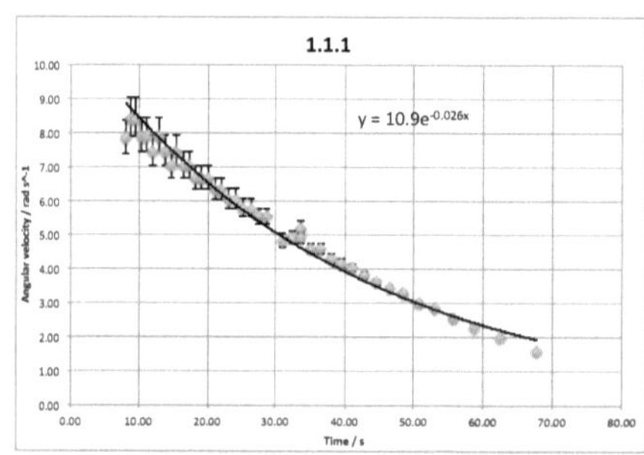


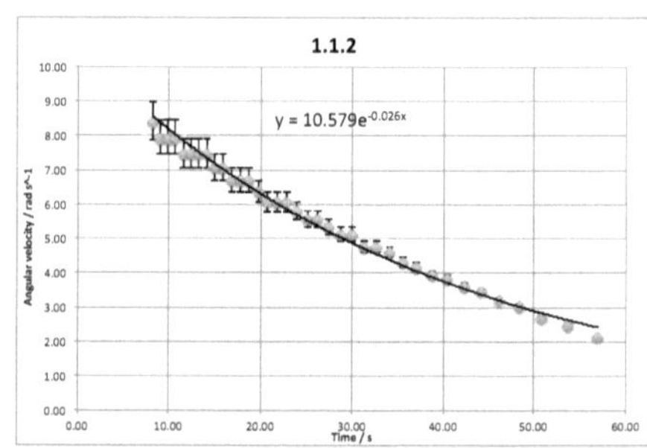


Diagram 10, 11: Experiments 1.1.1 and 1.1.2

The above are experiments 1.1.1 and 1.1.2. The first 1 means the p.d. across the magnet was ~5V, the second 1 means it was accelerated for 10s.

The error bars in the experiments 1.2.1 and 1.2.2 are clearly larger than those in the preceding experiments. This is because a difference in a smaller period has a greater effect on the angular velocity than the same effect in large period, as the percentage uncertainty is greater in the smaller and angular velocity is inversely proportional to the period.

Also, an effect caused by low sampling rate of the light barriers can be observed. At the high speeds in the beginning of the measurement, it can be seen that the jumps in angular velocity are discrete rather than continuous; plateaus form which partly even exist simultaneously. This is because the difference from one to the other is only a deviation of 0.05s in the period which is equal the sampling period. Values in between can hence not be recognised and they can exist simultaneously because even though the disc rotates at a momentarily constant speed, the light gates will maybe just catch the end of the cardboard along the rim of the disc. This effect will occur in all of the data. My sampling frequency was still reasonable though; the shortest period in this experiment was 0.4s, so the sampling frequency is still 8 times higher than the signal frequency and the Shannon-Theorem introduced on p. 5 only requires at least twice the signal frequency for a good sample.

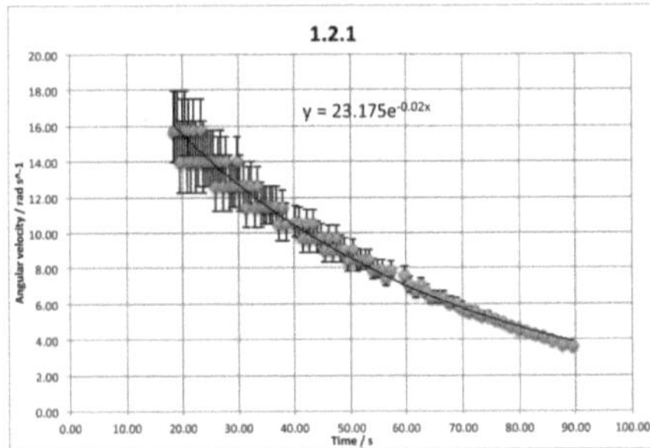
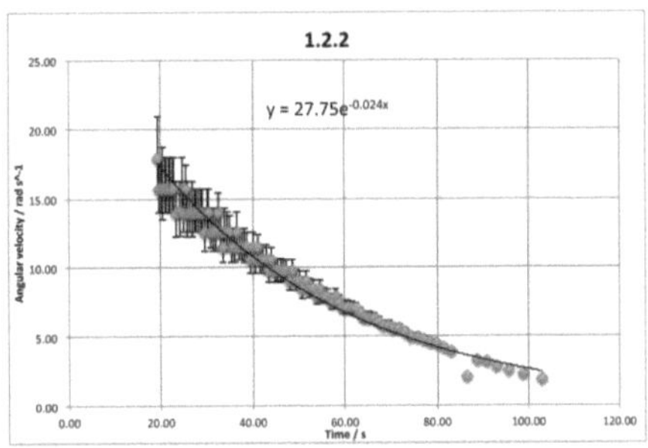

Diagram 12, 13: Experiments 1.2.1 and 1.2.2

In the experiments 1.3.1 and 1.3.2, I had to decrease the sampling period from 0.05s to 0.1s because it took longer for the disc to slow down in this experiment than in any of the others as I used the maximum acceleration and the minimum magnetic field strength.

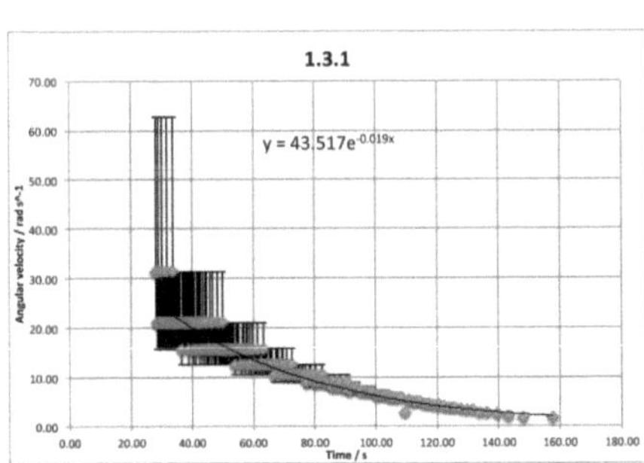
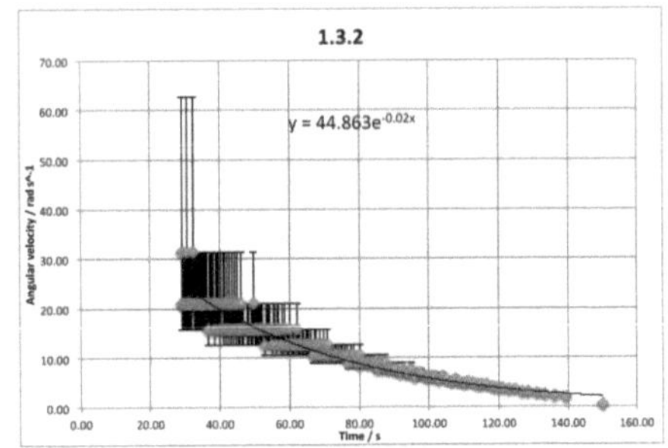

Diagram 14, 15: Experiments 1.3.1 and 1.3.2

The following diagrams are from the second series when the electromagnet was at ~10V. The initial speed increases down the page.

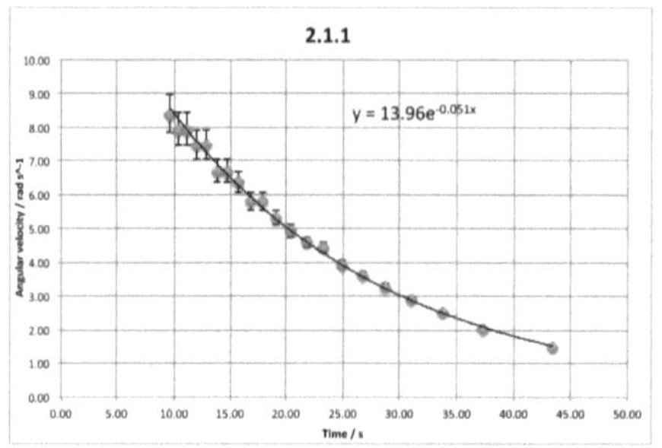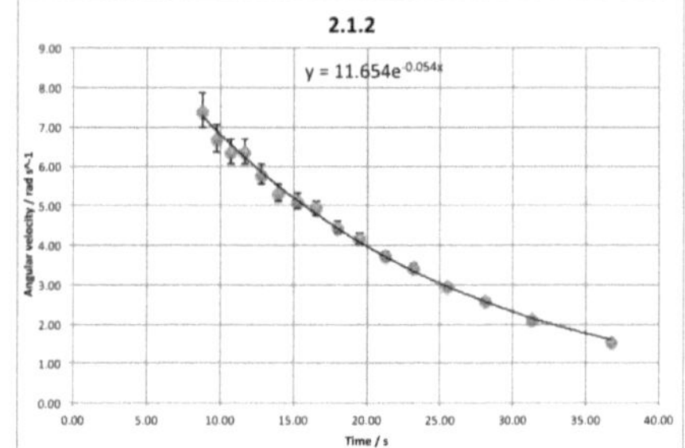

Diagram 16, 17: Experiments 2.1.1 and 2.1.2

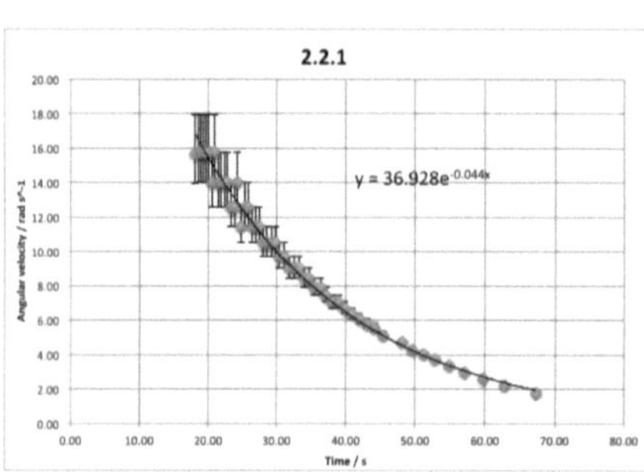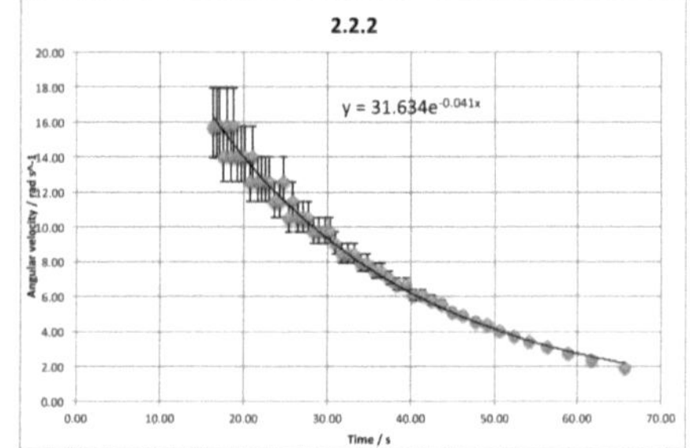

Diagram 18, 19: Experiments 2.2.1 and 2.2.2

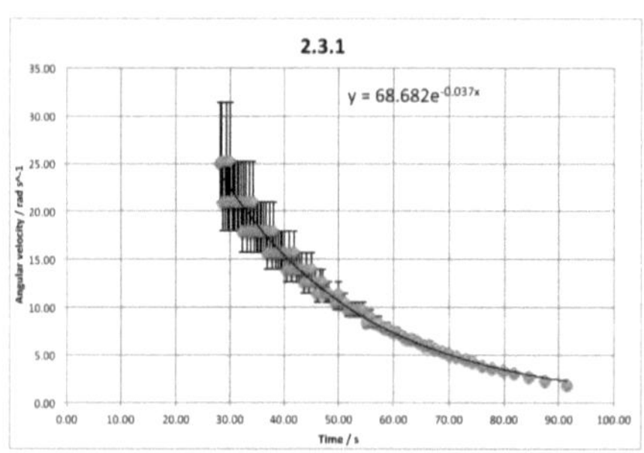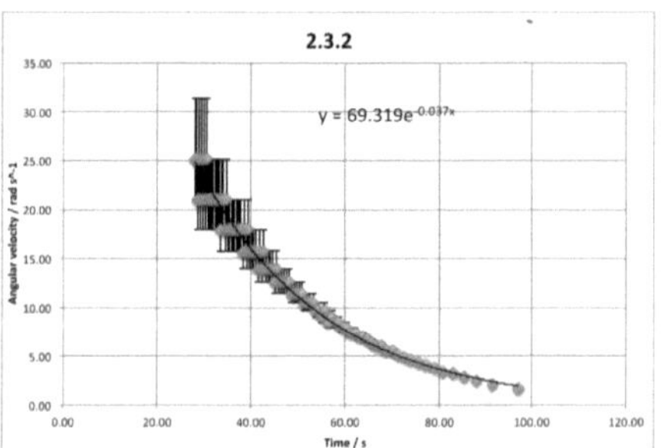

Diagram 20, 21: Experiments 2.3.1 and 2.3.2

In the above charts, it can be clearly observed how the magnitude of the deceleration increases with increasing initial speeds. This is because the emf induced is directly proportional to the rate of change of flux which is itself proportional to the angular velocity.

This is the last series, where the potential difference across the magnet was ~15V.

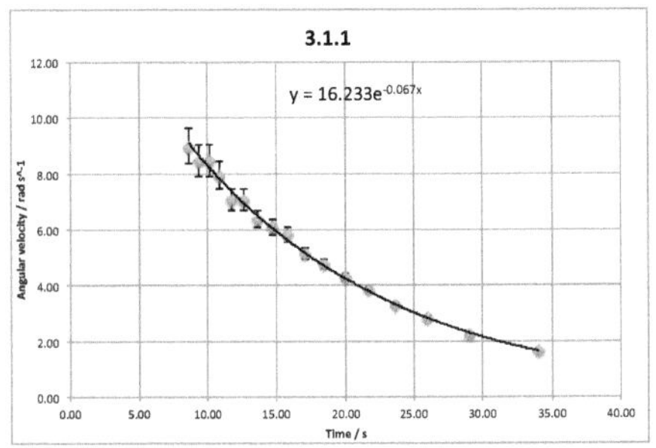


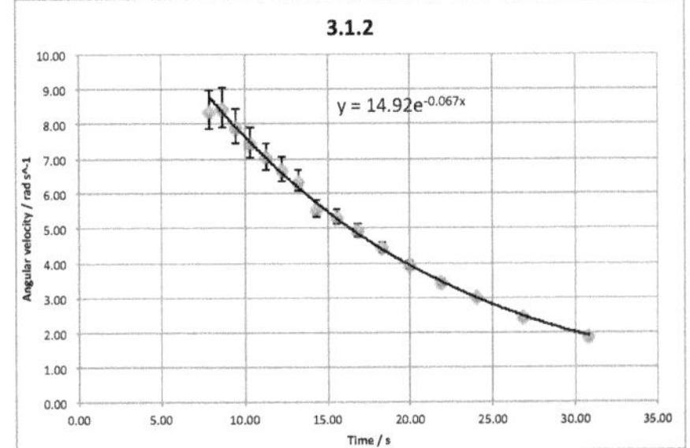


Diagram 22, 23: Experiments 3.1.1 and 3.1.2

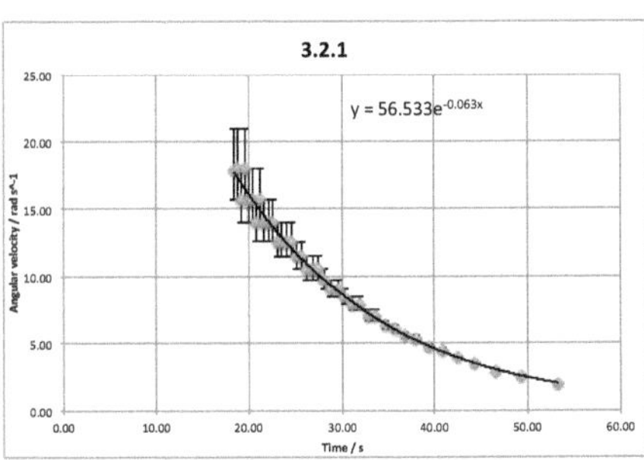


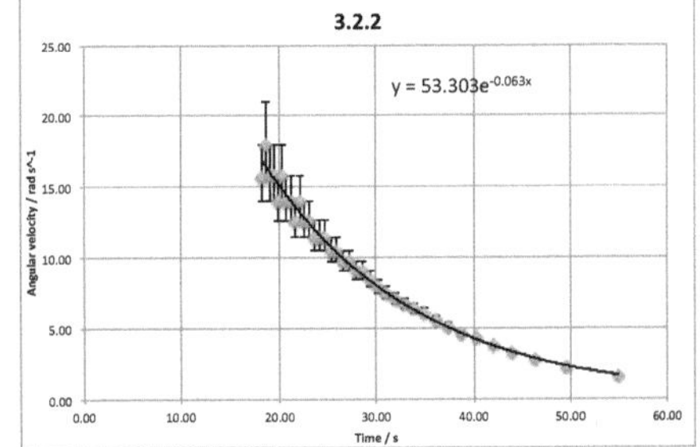


Diagram 24, 25: Experiments 3.2.1 and 3.2.2

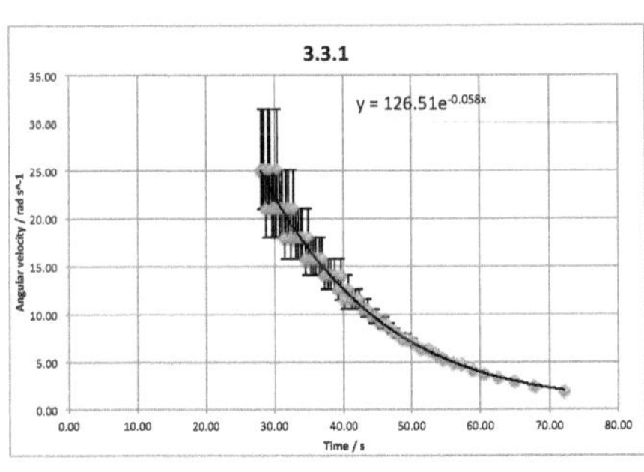


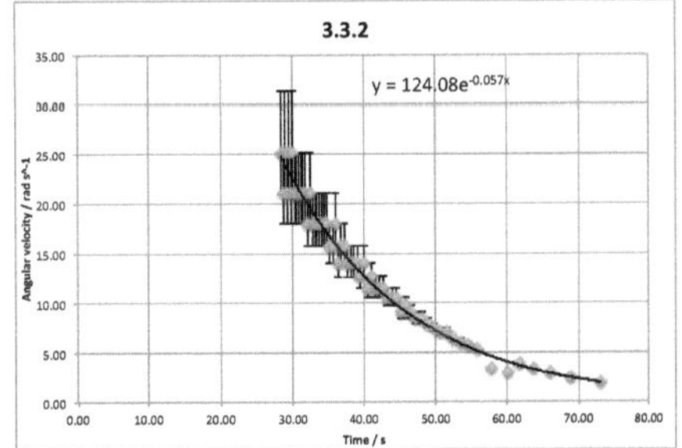


Diagram 26, 27: Experiments 3.3.1 and 3.3.2

To be able to quote actual initial velocities rather than the acceleration time (10s, 20s etc.), I have averaged all the initial velocities (in bold in the bottom row):

| 10 s | 20 s | 30 s |
| --- | --- | --- |
| 7.85 | 15.71 | 26.42 |
| 8.38 | 17.95 | 24.94 |
| 8.38 | 15.71 | 25.13 |
| 7.39 | 15.71 | 25.13 |
| 8.38 | 15.71 | 25.13 |
| 8.98 | 17.95 | 25.13 |
| **8.23 rad s⁻¹** | **16.46 rad s⁻¹** | **25.31 rad s⁻¹** |

These are the $\lambda$-values from all experiments. The units of the factors are s⁻¹ as an exponent has to be dimensionless and the factor is multiplied by a time.

| | 8.23 rad s⁻¹ | | 16.46 rad s⁻¹ | | 25.48 rad s⁻¹ | |
| --- | --- | --- | --- | --- | --- | --- |
| 5 V | 0.026 | 0.026 | 0.020 | 0.024 | 0.019 | 0.020 |
| 10 V | 0.051 | 0.054 | 0.044 | 0.041 | 0.037 | 0.037 |
| 15 V | 0.067 | 0.067 | 0.063 | 0.063 | 0.058 | 0.057 |

The values of the decaying factor have been plotted in the bar chart underneath with respect to the voltage across the magnet. The three different colours represent the different initial speeds in rad s⁻¹. An increase of the decaying factor with increasing strength of the magnet can be observed.

No error bars and uncertainties are included in any $\lambda$-values as they have been calculated by excel's regression and uncertainty or error cannot be carried through this easily.

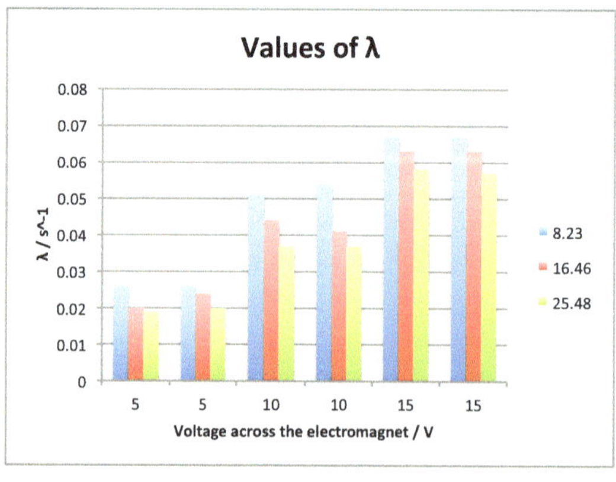


Diagram 28: Values of the decaying factors with respect to the p.d. across the electromagnet.

## 4.5 Acceleration Experiment

In this experiment, I measured the angular velocity in RPM of the motorised disc and the disc with the magnets attached. These two are plotted in the charts below. On the diagram, error bars are actually included, but as the laser pistol read to one decimal place so small that they can't be seen. The difference in velocity between the two discs is greater when the speed of the motorised disc is greater.

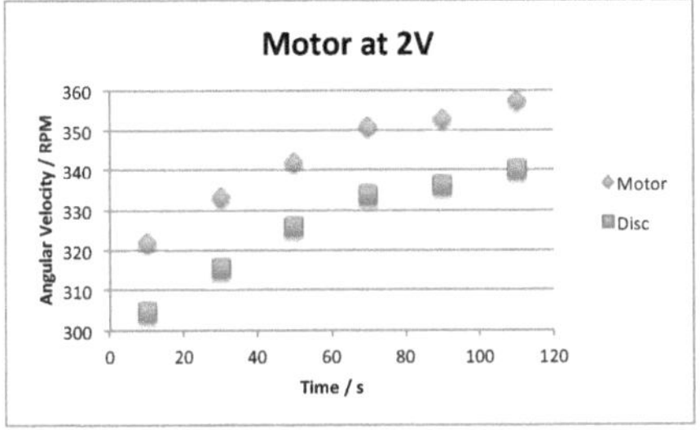
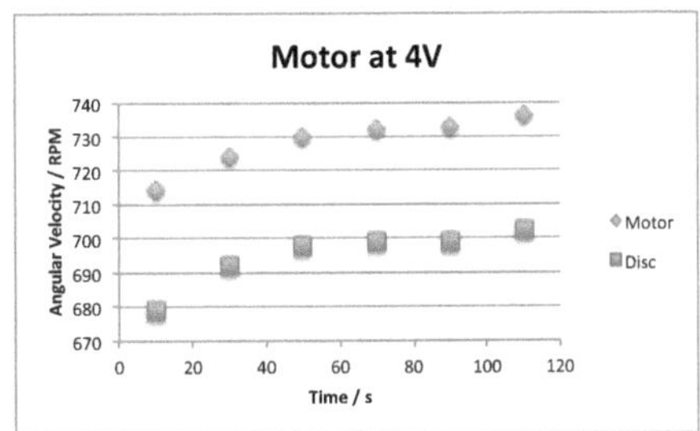

Diagram 29, 30: The velocities of the motorised and the dependent disc are shown independently on the two charts.

I then calculated the difference between the two and plotted this with respect to the time:

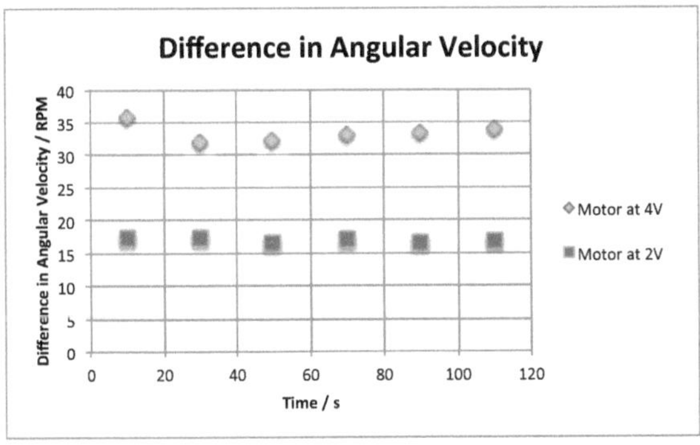

Diagram 31: Difference in angular velocity between the motorised and the dependent disc with respect to time.

The difference is reasonably constant, and as already outlined before, larger at higher speeds.

This experiment was primarily done in order to try wether the set-up worked at all, so not much interpretation will be given on these results. It would be very interesting to see, how the discs also behave with different magnetic field strengths, but fixing an electromagnet to the disc was here impossible for obvious practical reasons.

# 5. Interpretation

## 5.1 Discussion of uncertainty

**Bias and imprecise.** There is large uncertainty and large systematic error.

**Bias and precise.** There is small uncertainty but large systematic error.

**No bias but imprecise.** There is large uncertainty but small systematic error.

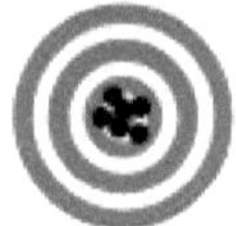

**No bias and precise.** There is small uncertainty and small systematic error. This result is accurate.

Figure 25: Precision and bias.

Results are **precise** if repeat measurements have values that are close together. Precise measurements have small random error and small uncertainty.

**Bias** arises from systematic errors which affect all the measurements in the same way, making them all higher or lower than the true value. Systematic errors do not average out.

Results are **accurate** if the are precise and free from bias.

In this experiment, the following factors caused imprecision and bias:

1. **Constant systematic bias** is for example a zero error in a measuring instrument. Constant systematic errors are difficult to remove, because their effects can only be observed if a way is found to remove them. Here, I did not notice any of these although they could have easily occured in on of the voltmeters or so.

2. **Varying systematic bias** occurs if the behaviour of an instrument changes with time or if an outside influence changes. In the main experiment, the coils in the electromagnet were heating up over the course of each measurement. This resulted in a change in voltage:

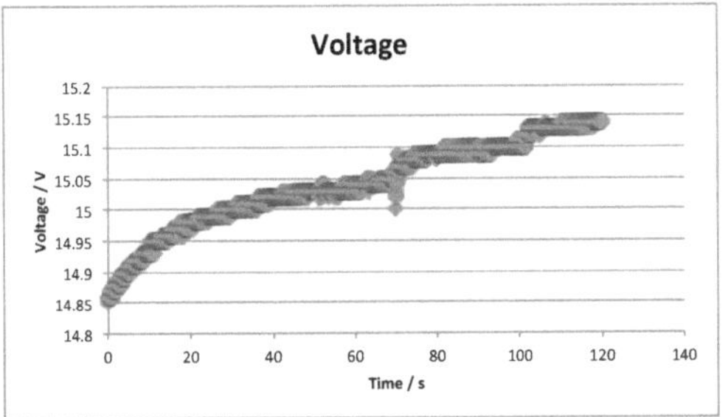

Diagram 32: Voltage with respect to time in a measurement with ~15 V.

The voltage is always in the range of 15.00 ± 0.15 V. I hence quoted ± 0.20 V for voltages ~15 V. I have done the same thing for the other to operating voltages (see p. 8).

3. **Limited resolution of an instrument** allows readings only to a certain change in input. This was the main source of uncertainty in this experiment as I have already outlined on pp. 14 and 18. It has been represented in error bars but also resulted in jumps in the velocity.

4. **Accidental momentary effects** are caused by some untoward event and produce outliers (anomalies). They can often be identified if a result is very different from others of if they depart a general trend. This does not seem to have happened often in my measurements, only in experiment 1.3.2, there is a possible outlier at t = 44.5s (circled in red). I subsequently deleted this value in order not to alter the regression line obtained. The corrected graph is displayed in the result section.

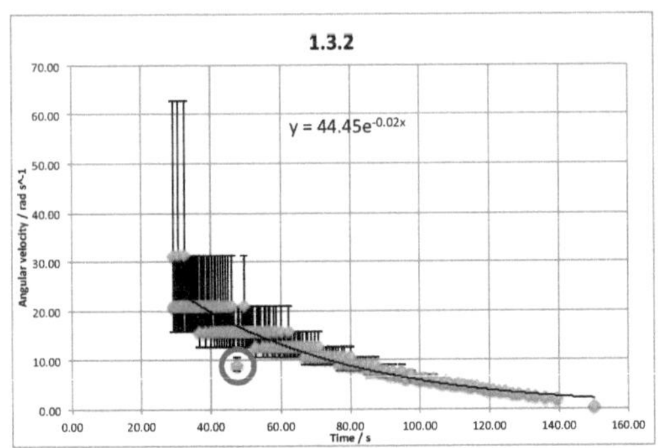

Diagram 33: Voltage with respect to time in a measurement with ~15 V.

Furthermore, I disconnected the motor from the disc and switched on the magnet at the instant when the stop-watch indicated the time to do so (either 10, 20 or 30 seconds) and there was of course a lot of imprecision in this process. This can be seen in the table on p. 21 which shows the initial velocities for each measurement.

## 5.2 Conclusion

Firstly, I would like to talk about the fact that the relationship between the angular velocity and the time is exponential. I managed to prove this by setting up a differential equation:

$$\frac{d^2\theta}{dt^2} \propto F \propto B \propto I \propto \varepsilon \propto -\frac{d\Phi}{dt} \propto -\frac{d\theta}{dt}$$

The angular deceleration of the disc is proportional to the force acting on the disc as **F = ma**. The force again is proportional to the magnetic field of the induced current in the disc and this magnetic field strength is proportional to the current in the disc. This current is driven by the electromotive force $\varepsilon$ and is proportional to it. As mentioned in the introduction, Faraday's law of electromagnetic induction states that the emf induced equals minus the rate of change of flux which is proportional to the angular velocity of the disc. Therefore,

$$\frac{d^2\theta}{dt^2} \propto -\frac{d\theta}{dt} \quad \text{so} \quad \frac{d^2\theta}{dt^2} + \lambda\frac{d\theta}{dt} = 0$$

Setting up an auxiliary equation: $m^2 + \lambda m = 0$, so $m = 0$ or $m = -\lambda$ and hence,

$$\theta = \alpha + \beta e^{-\lambda t}$$

Differentiating this:

$$\omega = \frac{d\theta}{dt} = -\lambda\beta e^{-\lambda t} = Ae^{-\lambda t}$$

This now proves that because the angular deceleration is proportional to the negated angular velocity, the angular velocity must decay exponentially.

In order to analyse the behavior of $\lambda$ with increasing magnetic field strength, I have created the following three graphs for each initial velocity:

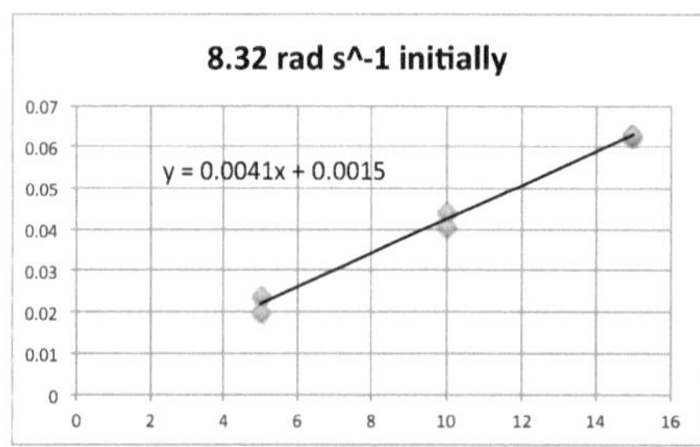


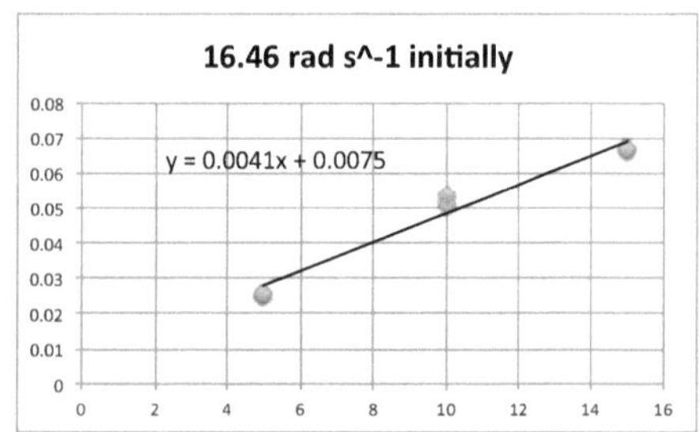


Diagram 34, 35: $\lambda$ with respect to the voltage across the electromagnet.

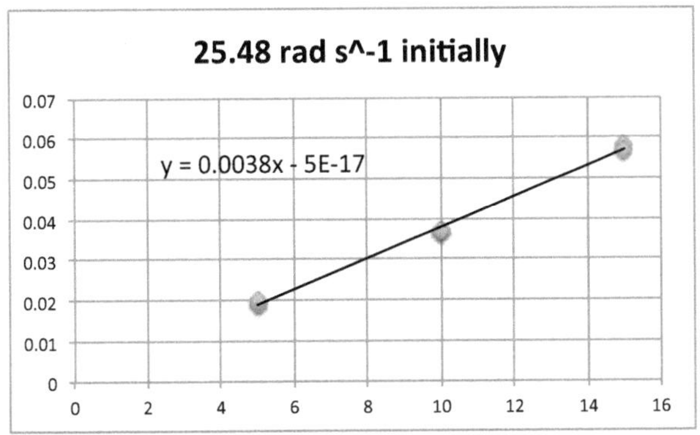


Diagram 36: $\lambda$ with respect to the voltage across the electromagnet.

The three curves have a very linear shape and their regression lines have very similar gradients: 0.0041, 0.0041 and 0.0038. Also, the values for   of each repeat lie extremely close together. This suggests, that the exponential decaying factor $\lambda$ in the amplitude of the angular velocity in eddy current breaking is directly proportional to the strength of the magnetic field inducing the emf that drives the eddy currents.

Wether or not the gradients will be similar for all initial velocities can not be determined from this investigation, too little data has been taken in order to make a judgement on this.

## REFERENCES

*An Introduction to the Sampling Theorem,* http://www.ti.com/lit/an/snaa079c/snaa079c.pdf.

Dobson, Grace, Lovett: *Physics,* Collins Advanced Science, 1997

Fullick, Patrick: *Physics,* Heinemann Advanced Science, 2003

Akrill, T & Bennet, G & Millar, C: *Physics,* Hodder & Stonyllton, 1994

Adams, Steve & Allday, Jonathan: *Advanced Physics,* Oxford University Press, 2000

*Dictionary of Physics,* Oxford University Press, 2005

## PICTURE CREDITS

All pictures and diagrams by David Brückner.